CLEAN AIR

Jill Sherman

Enslow Publishing
101 W. 23rd Street
Suite 240
New York, NY 10011
USA
enslow.com

Words to Know

atmosphere The mass of air that surrounds Earth.

carbon dioxide A gas that we breathe out.

evaporate When a liquid turns to a gas.

natural resource Something from nature that people use.

nitrogen A gas in the air that plants need to grow.

oxygen A gas in the air that we need to live.

pollution Dirt and unwanted chemicals in our air, soil, and water.

CONTENTS

The Air Around You

You can't see it. You can't touch it. But air is all around you. This invisible **natural resource** has been hiding right under your nose!

FAST FACT
Air rises far above Earth's surface, all the way to outer space.

Giving Life

Take a deep breath. Your lungs fill with air. Everyone needs air to stay alive. Air is a mixture of gases that animals and plants use.

Fast Fact

Air is mostly nitrogen and oxygen, but many other gases make up Earth's air.

O_2
CO_2

The Oxygen Cycle

We use Earth's air. People and animals breathe in oxygen. They breathe out carbon dioxide. Plants take in carbon dioxide. They give off oxygen. The gases in the air are always moving. It is called a cycle.

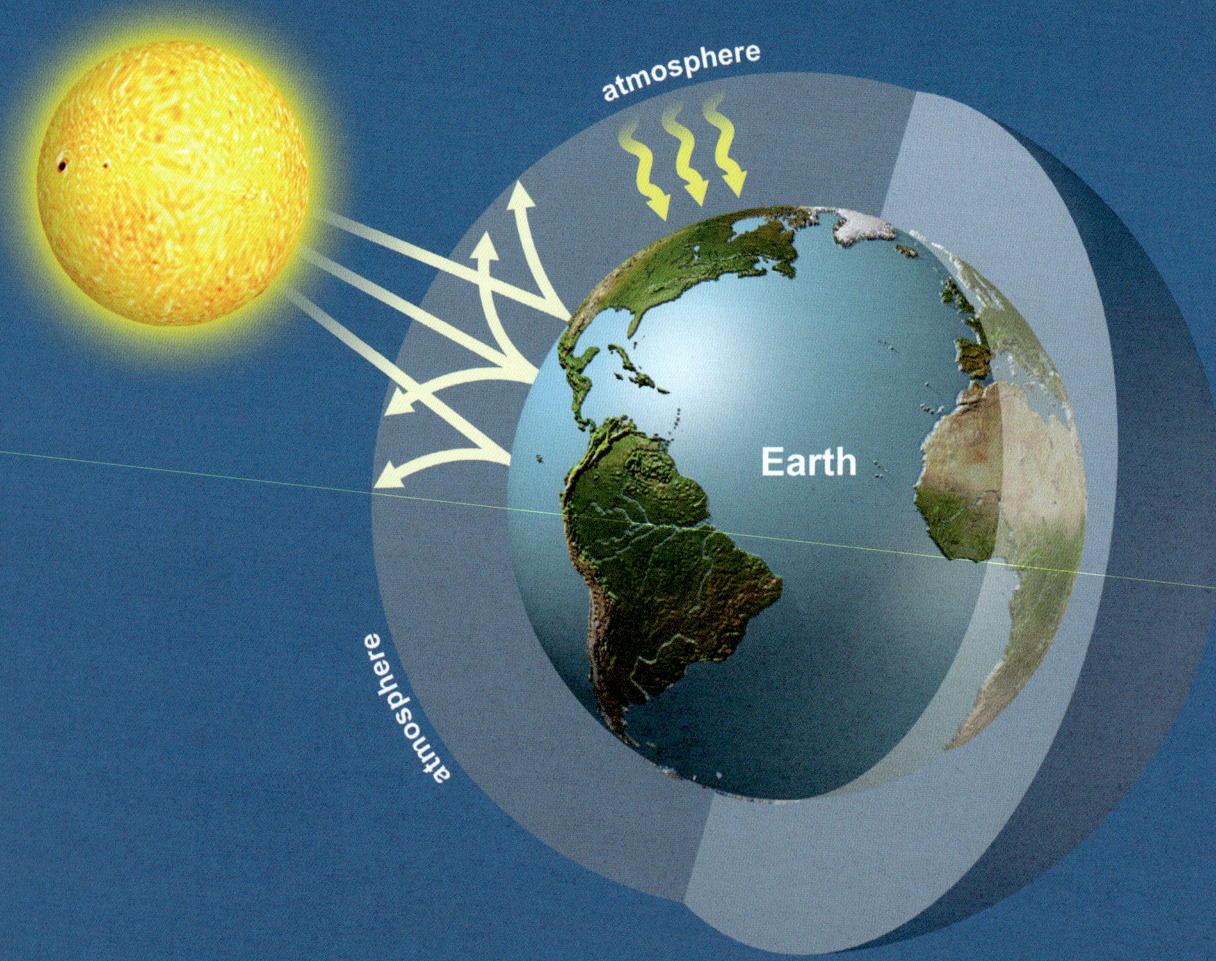
Sun
atmosphere
Earth
atmosphere

What Air Does

Air does more than let us breathe. It surrounds Earth, making up the **atmosphere**. Air holds in the sun's warmth. It also protects from damage the sun could cause.

Fast Fact

Without an atmosphere, Earth would not have water. It would **evaporate** into space.

Moving Air

Outside, the wind blows against your cheek. Wind is just moving air. But when the wind blows, it carries things with it. Seeds may travel far distances on the wind.

Fast Fact

A sunny day may change to a sudden storm. Moving air creates our changing weather.

Wind Energy

Wind is powerful. We can harness the wind's power to do great things. Sails capture the wind and move ships. When wind moves the blades of a windmill, we can make electricity.

Fill It Up

Humans have found other ways to use air. Can you bounce a deflated basketball? Or ride on flat tires? No! But fill them with air, and you are ready to roll.

Let It Out

Let air out of a balloon and it blows. People use canned air to clean things. Release the air to remove dust from your computer.

FAST FACT
Fans also make air move. Cool down a warm room by moving air with a fan.

Clean Air

Pollution sickens the air. If it gets into your lungs, you cough. If it fills the sky, it blocks light from reaching plants. Keep Earth's air clean and it will continue to take care of us.

FAST FACT

Pollution can be natural, coming from disasters like forest fires or volcanoes. Or it can be manmade, coming from factories and automobiles.

Activity

Particles in the Air

Materials
2 index cards
petroleum jelly
tape
magnifying glass

Procedure:

1. Smear a small amount of petroleum jelly on each index card.

2. Place one card indoors and the other outdoors. Tape each card down so that they stay put.

3. Let the cards sit for a few days.

4. Examine each card using your magnifying glass. What do you see in the jelly? Which area carries more particles in the air?

Learn More

Books

Ajmera, Maya, and Dominique Browning. *Every Breath We Take*. Watertown, MA: Charlesbridge, 2016.

MacAulay, Kelley. *Why Do We Need Air?* New York, NY: Crabtree Publishing. 2014.

Nunn, Daniel. *Why Living Things Need Air*. New York, NY: Heinemann Library, 2014.

Websites

NASA: Climate Kids
climatekids.nasa.gov/10-things-air/
NASA's Eyes on the Earth presents ten things to learn about air.

National Center for Atmospheric Research: Kids' Crossing
eo.ucar.edu/kids/sky/air1.htm
Check out games, links, and fun facts about air.

Index

Published in 2018 by Enslow Publishing, LLC.
101 W. 23rd Street, Suite 240, New York, NY 10011

Library of Congress Cataloging-in-Publication Data

Names: Sherman, Jill.
Title: Clean air / Jill Sherman.
Description: New York : Enslow Publishing, 2018. | Series: Let's learn about natural resources | Audience: K to grade 3. | Includes bibliographical references and index.
Identifiers: LCCN 2017011015| ISBN 9780766091450 (library bound) | ISBN 9780766091436 (pbk.) | ISBN 9780766091443 (6 pack)
Subjects: LCSH: Air—Juvenile literature. | Atmosphere—Juvenile literature.
Classification: LCC QC161.2 .S527 2018 | DDC 551.5—dc23

LC record available at https://lccn.loc.gov/2017011015

Printed in China

Photo Credits: Cover, p. 1 Hans-Peter Merten/Photolibrary/Getty Images; interior pages (soil, grass, sky) Andrey_Kuzmin/Shutterstock.com; interior pages (sign) johavel/Shutterstock.com; p. 4 Inara Prusakova/Shutterstock.com; p. 6 CebotariN/Shutterstock.com; p. 8 Sergey Merkulov/Shutterstock.com; p. 10 Vitoriano Junior/Shutterstock.com; p. 12 Photobank gallery/Shutterstock.com; p. 14 fokke baarssen/Shutterstock.com; p. 16 Sergey Novikov/Shutterstock.com; p. 18 hanapon1002/Shutterstock.com; p. 20 Igor Iakovlev/Shutterstock.com.